BIG AND SMALL ANIMALS

MAMMALS

BY BRENNA MALONEY

Children's Press®
An imprint of Scholastic Inc.

A special thank-you to the team at the Cincinnati Zoo & Botanical Garden for their expert consultation.

Library of Congress Cataloging-in-Publication Data
Names: Maloney, Brenna, author.
Title: Mammals / by Brenna Maloney.
Description: First edition. | New York, NY: Children's Press, an imprint of Scholastic Inc., 2023. | Series: Big and small animals | Includes index. | Audience: Ages 5–7. | Audience: Grades K–1. | Summary: "Continuation of the Wild World series comparing big and small animal sizes"— Provided by publisher.
Identifiers: LCCN 2022026384 (print) | LCCN 2022026385 (ebook) | ISBN 9781338853568 (library binding) | ISBN 9781338853575 (paperback) | ISBN 9781338853582 (ebk)
Subjects: LCSH: Mammals—Miscellanea—Juvenile literature. | CYAC: Mammals.
| Size. | BISAC: JUVENILE NONFICTION / Animals / Mammals | JUVENILE NONFICTION / Concepts / Opposites
Classification: LCC QL706.2 .M25 2023 (print) | LCC QL706.2 (ebook) | DDC 599—dc23/eng/20220606
LC record available at https://lccn.loc.gov/2022026384
LC ebook record available at https://lccn.loc.gov/2022026385

Copyright © 2023 by Scholastic Inc.

All rights reserved. Published by Children's Press, an imprint of Scholastic Inc., *Publishers since 1920*. SCHOLASTIC, CHILDREN'S PRESS, and associated logos are trademarks and/or registered trademarks of Scholastic Inc.

The publisher does not have any control over and does not assume any responsibility for author or third-party websites or their content.

No part of this publication may be reproduced, stored in a retrieval system, or transmitted in any form or by any means, electronic, mechanical, photocopying, recording, or otherwise, without written permission of the publisher. For information regarding permission, write to Scholastic Inc., Attention: Permissions Department, 557 Broadway, New York, NY 10012.

10 9 8 7 6 5 4 3 2 1   23 24 25 26 27

Printed in China   62
First edition, 2023

Book design by Kay Petronio

Photos ©: cover top, back cover right, 1 top, 2 right: Phillip Colla/BluePlanetArchive; cover bottom, back cover left, 1 bottom, 2 left: MerlinTuttle.Org; 4 top right: MerlinTuttle.Org; 5 center: SCIEPRO/Science Source; 6: MerlinTuttle.org/Science Source; 7 right: MerlinTuttle.Org; 8–9: MerlinTuttle.Org; 10–11: Science History Images/Alamy Images; 12–13: Christophe Courteau/NPL/Minden Pictures; 13 inset: Michel & Christine Denis-Huot/Biosphoto; 14–15: Steve Newbold/Getty Images; 16–17: Jurgen & Christine Sohns/FLPA/Minden Pictures; 18–19: Penny Boyd/Alamy Images; 26 top: Christopher Swann/BluePlanetArchive; 27: Flip Nicklin/Minden Pictures; 28–29: Franco Banfi/BluePlanetArchive; 30 top left: SCIEPRO/Science Source; 30 top right: MerlinTuttle.Org; 30 bottom right: shot by supervliegzus/Getty Images.
All other photos © Shutterstock.

↑ KITTI'S HOG-NOSED BAT

↑ BLUE WHALE

CONTENTS

Mammal Matters 4
#10 Smallest Mammal:
   Kitti's Hog-Nosed Bat 6
Kitti's Hog-Nosed Bat Close-Up 8
#9: Moonrat 10
#8: Giant Anteater 12
#7: Capybara 14
#6: Red Kangaroo 16
#5: Giant Forest Hog 18
#4: West Indian Manatee 20
#3: Southern Elephant Seal 22
#2: African Elephant 24
#1 Biggest Mammal: Blue Whale ...26
Blue Whale Close-Up 28
Mammals Big and Small 30
Glossary 31
Index 32

MAMMAL MATTERS

Humans, elephants, whales, and kangaroos are all **mammals**! All mammals have **backbones**. All mammals have lungs, and they breathe air. All mammals have hair or fur. Most mammals give birth to live young. Mammal mothers nurse their young with milk. Mammals are also **warm-blooded**. This means their body temperature does not change with their surroundings.

FACT: Mammals have another thing in common: They all have three tiny bones in their middle ears.

Measuring Up

This book is all about size. Which mammals are the smallest? Which are the biggest? Why does being big or small matter? An animal's size can determine where and how it lives. Size makes a difference as to what **prey** it can chase and what **predators** it must flee from. You can learn a lot about mammals just by their size.

Get ready to discover the different sizes of 10 fantastic mammals and why it matters, from the smallest to the biggest!

#10 Smallest Mammal: KITTI'S HOG-NOSED BAT

Kitti's hog-nosed bats are so small, they're often called bumblebee bats. These 1.1-inch-long (28 mm) bats are the smallest mammals on Earth *by length*. They weigh only .07 ounces (2 grams). They live in **roosts** inside caves, far from the cave's entrance. Clustered in small groups of about 100, these bats don't often stray more than half a mile (1 km) from their roost site.

Kitti's hog-nosed bats only look for food during short windows of time. They search for about 30 minutes in the evening and 20 minutes before dawn. These bats are **insectivores**. They hunt insects using **echolocation**. They make squeaking sounds that echo. The sound bounces off their surroundings, giving the bats an accurate location of their prey. They mostly eat small flies, but they also eat spiders, beetles, wasps, and lice.

SAME SIZE AS . . .
A Kitti's hog-nosed bat weighs as much as two Skittles candies.

FACT: This bat can be found only in Thailand and Myanmar.

KITTI'S HOG-NOSED BAT CLOSE-UP

Kitti's hog-nosed bat is not just the smallest mammal by length. It is also the smallest bat on the planet.

WINGS
Long tips on this bat's wings allow it to hover.

COAT
Unlike most other bats, which have a black or brown coat, this bat has a reddish-brown or gray coat.

CLAWS
Strong claws allow this bat to hang upside down without using much energy.

EYES
This bat's eyes are small and mostly hidden by fur.

EARS
Its ears are relatively large, given the size of its head.

SNOUT
As its name suggests, its snout is similar to a hog's.

TEETH
Front teeth do most of the work, snatching and gripping insects during flight.

LEGS
While its legs are particularly slender, there is a large web of skin between the legs that may help this bat fly.

FACT: Kitti's hog-nosed bat was discovered by Kitti Thonglongya, a scientist, in 1973.

#9 MOONRAT

FACT: Moonrats are **nocturnal**, meaning they are active at night.

Believe it or not, moonrats are not rats. They are the world's largest hedgehogs. They can weigh as much as 3 pounds (1 kg). Using their teeth and long snouts, these **omnivores** scratch and poke rotten trunks and leaf litter to find worms and insects to eat. They also eat fruit, and sometimes tiny frogs or fish. Moonrats are **solitary** animals. They have a strong, stinky smell to keep others away.

SAME SIZE AS . . .

This hedgehog can weigh as much as a toaster.

SAME SIZE AS . . .

A giant anteater weighs more than three bowling balls.

FACT

This anteater's tongue can be 2 feet (0.6 m) long.

#8
GIANT ANTEATER

The average weight of a giant anteater is about 60 pounds (27 kg). How do they get so big? Giant anteaters have no teeth, but they don't need them! Their long tongues can lap up as many as 35,000 ants and termites every day. They use sharp claws to dig into termite and ant mounds. The insects scurry away, but not fast enough.

#7 CAPYBARA

FACT: These mammals live in Central and South America.

Capybaras are the world's largest rodents. They are closely related to guinea pigs. They can weigh as much as 150 pounds (68 kg). Capybaras are found in areas where water can easily be found. Like other rodents, capybaras' teeth never stop growing. They wear them down by chomping on plants and grasses. Capybaras have partially webbed toes that help them paddle around in water as they search for food.

SAME SIZE AS...

A capybara can weigh as much as three large bags of dog food.

SAME SIZE AS . . .

A red kangaroo can weigh as much as a 6-foot-tall (2 m) man.

#6 RED KANGAROO

The largest of all kangaroos, red kangaroos can weigh up to 200 pounds (91 kg). These **herbivores** have long and powerful back legs and feet. They move around in open grassland and desert **habitats**. They can move at great speeds—more than 35 miles per hour (56 kph)—and can cover great distances. They eat grasses and other plants. If water is hard to find, they get water from the plants they eat.

FACT
Male red kangaroos are larger than female red kangaroos.

#5 GIANT FOREST HOG

Giant forest hogs are the largest members of the wild pig family. They can weigh up to 600 pounds (272 kg). The giant forest hog is mainly an herbivore, but it will also eat dead animals. Mostly active at dusk and dawn, forest hogs look for grasses, leaves, and other plants to eat. Forest hogs use their short tusks and teeth to dig and uproot plants.

FACT: This hog is native to Africa.

SAME SIZE AS . . .

A giant forest hog can weigh as much as a grizzly bear.

SAME SIZE AS...

A West Indian manatee can weigh about as much as five full-size refrigerators.

20

FACT: Manatees sometimes go by the nickname "sea cows."

#4 WEST INDIAN MANATEE

West Indian manatees are not small mammals. They can weigh up to 1,200 pounds (544 kg). Despite their size, West Indian manatees can move easily in water. They can swim upside down and do rolls and somersaults! Manatees eat **aquatic** plants, such as cordgrass, turtle grass, and eelgrass. Manatees have to eat a lot each day to be full. They might spend eight hours a day eating.

#3 SOUTHERN ELEPHANT SEAL

Southern elephant seals are the largest of all seals. Males can be more than 20 feet (6 m) long and weigh up to 8,800 pounds (3,992 kg). Southern elephant seals are open-ocean predators and spend much of their time at sea. They have been recorded diving up to 7,000 feet (2,134 m) and can stay underwater for nearly two hours. Using their sharp teeth, southern elephant seals can dig through mud or snatch prey out of the water. As **carnivores**, they eat squid, mollusks, teeny tiny shrimp-like animals called **krill**, and algae.

These seals live in sub-Antarctic and Antarctic waters.
FACT

SAME SIZE AS...
A southern elephant seal weighs as much as a flatbed truck.

23

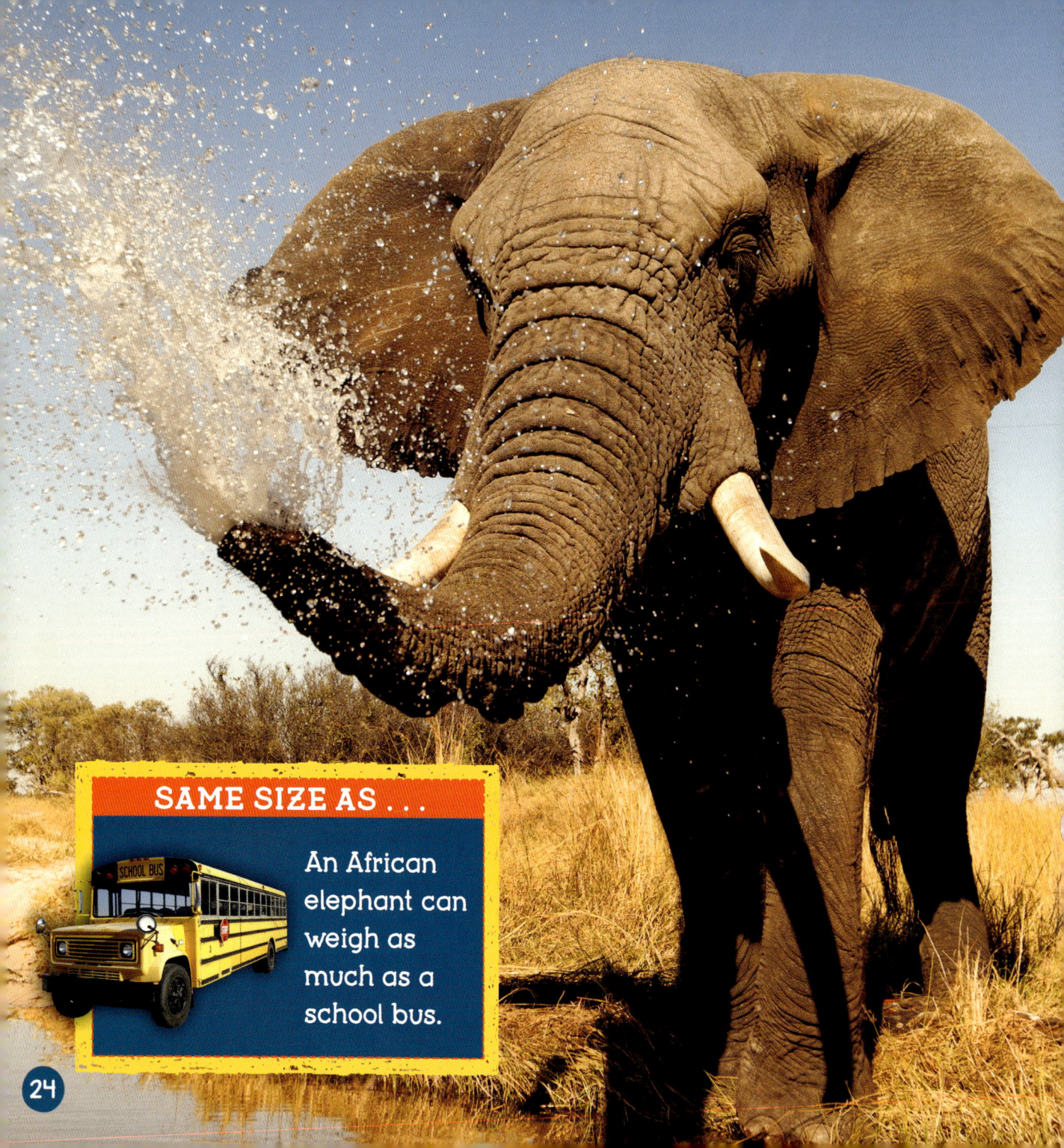

SAME SIZE AS . . .

An African elephant can weigh as much as a school bus.

#2 AFRICAN ELEPHANT

African elephants are the largest land animals on Earth, weighing up to 7 tons (6,350 kg). They eat roots, grasses, fruit, and bark. An adult can eat up to 300 pounds (136 kg) of food per day. They can also drink as much as 50 gallons (189 liters) of water in a single day. Their long, flexible trunks help them smell, drink, trumpet, and grab objects. African elephants can even use their trunks as snorkels when swimming.

FACT: There are two **species** of African elephants—savanna and forest. Savanna elephants are larger.

#1 Biggest Mammal: BLUE WHALE

SAME SIZE AS . . .
A blue whale weighs as much as a 300-seat Boeing 787.

Hands down, blue whales are the largest mammals on Earth. They are also the largest *animals* on Earth. They are, in fact, the largest animals to have ever lived on our planet! Blue whales can weigh up to 200 tons (181,437 kg). What do these enormous animals eat? When blue whales hunt for food, they swim into large **schools** of krill with their mouths open, also taking in giant mouthfuls of water.

Blue whales have stiff, fringed plates—called **baleen**—attached to their upper jaws. When they close their mouths, they trap the water inside. When they press their tongues down, the water seeps out through the baleen. Thousands of krill are left behind and swallowed. Blue whale stomachs can hold 2,200 pounds (998 kg) of krill at a time.

FACT: Blue whales are found in all oceans except the Arctic.

HEAD
Its broad, flat head appears U-shaped from above.

BLOWHOLES
A tall, column-like spray can be seen when the blue whale surfaces and pushes out air from its two blowholes.

TONGUE
The tongue forces water out through the thin, overlapping baleen plates.

THROAT
A blue whale's throat can stretch wider as it takes in water.

MOUTH
A series of fringed plates—baleen—hang in the whale's mouth to strain seawater for food.

HEART
A blue whale's heart is the largest heart in the animal kingdom, weighing about 400 pounds (181 kg).

FLUKE
Wide, triangular **flukes** are attached to a thick tail.

BODY
Long and slender, a blue whale's body can be different shades of gray and blue.

BLUE WHALE CLOSE-UP

Blue whales are not only the largest animals, they are also the loudest animals on Earth. The songs of a blue whale can be heard from 1,000 miles (1,609 km) away.

FACT Blue whales can live to be 80 years old or older.

MAMMALS BIG AND SMALL

Mammals come in all shapes and sizes. Now you know why it matters to be a big or small mammal. There are more than 5,400 mammal species on Earth—too many to cover in this book! Make it your mission to learn even more about these amazing animals, both big and small.

GLOSSARY

aquatic (uh-KWAH-tik) living or growing in water

backbone (BAK-bohn) a set of connected bones that runs down the middle of the back; also called the spine

baleen (buh-LEEN) the bristly plates that filter, sift, sieve, or trap a whale's food

carnivore (KAHR-nuh-vor) an animal that eats other animals for food

echolocation (ek-oh-loh-KAY-shuhn) the ability of some animals, such as dolphins and bats, to locate objects by reflected sound

fluke (flook) one of the two flat, wide fins of a whale's tail

habitat (HAB-i-tat) the natural place where an animal or plant is usually found

herbivore (HUR-buh-vor) an animal that eat plants

insectivore (in-SEK-tuh-vor) an animal that feeds mainly on insects

krill (kril) a small, shrimp-like animal that lives in the ocean

mammal (MAM-uhl) a warm-blooded animal that has hair or fur and usually gives birth to live babies; female mammals produce milk to feed their young

nocturnal (nahk-TUR-nuhl) an animal that is active at night

omnivore (AHM-nuh-vor) an animal that eats both plants and animals

predator (PRED-uh-tur) an animal that lives by hunting other animals for food

prey (pray) an animal that is hunted by another animal for food

roost (roost) a place where bats come together to rest during the day

school (skool) a large group of fish or sea mammals

solitary (SAH-li-tair-ee) not requiring or without the companionship of others

species (SPEE-sheez) a group of similar organisms that are able to reproduce

warm-blooded (WORM bluhd-id) having a body temperature that does not change, even if the temperature of the surroundings is very hot or very cold

INDEX

Page numbers in **bold** indicate images.

A
African elephant, 5, **24–25**, 30
  diet and eating, 25
  species of, 25
anteater. *See* giant anteater

B
bat. *See* Kitti's hog-nosed bat
blue whale, 5, **26**, **27**, **28–29**, 30
  diet and eating, 26–27
  lifespan, 29
  sounds, 29
bumblebee bat. *See* Kitti's hog-nosed bat

C
capybara, **14–15**, 30
  diet and eating, 14
  habitat, 14

G
giant anteater, **4**, **12–13**
  diet and eating, 13
giant forest hog, **18–19**
  diet and eating, 18
  habitat, 18

K
Kitti's hog-nosed bat, **4**, **6**, **7**, **8–9**, 30
  diet and eating, 7
  habitat, 6, 7
  nickname, 6
  sounds, 7

M
mammals
  common traits of, 4
  number of, 30
moonrat, **10–11**, 30
  diet and eating, 10
  social habits, 10

R
red kangaroo, 5, **16–17**
  diet and eating, 17
  habitat, 17
  speed, 17

S
sea cow. *See* West Indian manatee
southern elephant seal, **22–23**
  diet and eating, 22
  habitat, 23

W
West Indian manatee, **20–21**
  diet and eating, 21
  nickname, 21
whale. *See* blue whale

ABOUT THE AUTHOR

Brenna Maloney is the author of more than a dozen books. She lives and works in Washington, DC, with her husband and two sons. She wishes she had more pages to tell you about mammals. She also wishes she had the empathy of an Asian elephant, the toughness of a Sumatran rhino, and the mustache of a bearded emperor tamarin. (Look them up!)